The Christian Religion

Twenty-eight unacceptable mysteries to the intelligent man or

Why doesn't God give a damn?

A monograph written by

Dennis R. Horner

seasidepress.org

amzn ppbck 978-1-947940-22-2

Introduction

The universe is as it is because there is no other way it can be. We have a universe, that is governed at each and every instant of time, by pure chance.

Originally everybody believed in God because his existence was the only way anyone could think of to explain the existence and behavior of the world - better perhaps to say the Earth. But slowly at first and now faster and faster, the unexplained phenomena (the existence and unexplained ability of which led to the belief in God in the first place) - were explained and so the reasons for believing in a God diminished. As for chaos, it is just that which convinces me that no God is needed - everything happens by pure happenstance. It is such an easy concept, and of course it is experimentally verifiable.

What evidence is there for His existence? Or, for that matter, for His non-existence? For I don't contend that He doesn't exist - what I say is that the loving God of the Christian/ Muslim mythology doesn't exist, and for that view I have ample evidence.

What is my evidence that God/Allah does not exist? (I'm not sure about Jehovah/Yahweh because as I understand it, the Jewish God isn't all that of a nice guy - apart from having the Jews as His chosen people, of course - according to the Jews that is, and my argument is that if He exists He just couldn't care less about us).

The Roman Church founded by St. Paul himself, has a mass of homosexuals, little-boy-buggery priests on its staff. So the Cardinal of Boston went off to the Pope (I don't mean recently, I mean some while ago) and they had a confab about what they ought to do about it, and they decided to hush it all up! Now you may say that they got it wrong - Pope and Cardinal, both are human. But hang about. Didn't they get down on their knees and have a good old pray to their God asking for guidance? Is it conceivable that they didn't have a prayer session? And Jesus said something like "When two or three are gathered together in my name…(God)… will grant their request. But He didn't, did He? He didn't give His chief minister here on Earth any guidance at all, did He, even though this omniscient God knew a wrong decision would bring the Pope and the Roman Church into such disrepute, that Roman Catholics are now quitting in droves resulting, of course, in their eternal damnation. God has quite deliberately failed to advise his Pope - indeed, if you believe in Him, you really ought to say that God has deliberately conspired with Pope and Cardinal to cause the damnation of millions of people. Go on, you explain it some other way - how the prayers for guidance of Pope and Cardinal remains unanswered.

We are nothing at all, just an odd occurrence on a random planet in a random galaxy in one of Heaven knows how many universes. Do you matter? Do I matter? Does anything that happens on the Earth matter? Of course not. The obvious explanation is that man, in his absurdly arrogant conceit, has constructed an enormous edifice of pure bull in an attempt to bolster his own importance.

Dennis R. Horner

***

THE CHRISTIAN RELIGION

As flies to wanton boys are we to the gods,
They kill us for their sport.
King Lear, Act IV, sc. 1

Philosophy

After a great deal of thought, I have come to the conclusion that the Christian religion is a load of rubbish. I must apologize for putting it so bluntly, but that is my view and that which I wish to expound here. Indeed, although I was brought up to be a Christian and so here I'm attacking this particular Faith, I suspect that most of what I say applies equally well, *mutatis mutandis*, to the other two monotheistic Faiths, Judaism and Islam, of both of which I know relatively little, and also to the other world religions of which I know even less. I have no quarrel with the ethics of these great Faiths; it is with what I would call the *superstition* that is grafted on to them, that my quarrel lies.

We must know not to confuse ethics with religion. The ethical teaching of most religions are fine, and I tried to obey them. In particular, the teachings of Jesus Christ are superb - but that seems to me to have nothing to do with the religious (I would say superstitious) elements told by the Church which are, in my view, wholly unjustified. For example, my deeply religious brother in law, a man of great integrity, once asked me why I am a relatively moral person when I deny the teachings of religion. I found the question surprising; I could not see the connection, because I accept the ethics and I deny the (unproven) superstition.

The first point I would make is that I do not deny the existence of a God in the sense of a Prime Mover, a Someone or Something that set things initially in motion, for it seems to me that whether or not such a God exists, must forever remain unknown and unknowable. I do not deny the existence of a God in that sense, but neither do I accept it. I just don't know.

However, I think it is necessary to dispose of the argument that the universe must have a cause. *Every thing must have a cause* is arrant nonsense, because it gives rise to an infinite regress. The cause must have a cause, and the cause of the cause must have a cause, and so on*, and anyway, modern physics teaches us that some things don't have causes (e.g. the decay of radioactive nuclei) so I am quite happy to postulate that the famous Big Bang had no cause and all the rest is happenstance. On the other hand, there may well be a Prime Mover. Who knows?

**I press the switch which causes the light to come on; I want the light to come on and that wish causes me to press this switch; I want to read a newspaper article and that causes me to want more light; there is an article about global warming; that causes me to want to read it... and so on and so on.*

Now had I been born a thousand years ago - had I been writing this in the year 1000 AD - there would have been countless reasons to believe in God and I am pretty sure I would have believed in Him. There were so many apparent miracles which were unexplained:

1. The sun and the moon traveled daily across the sky at different speeds.
2. The planets also moved, and now differently again.
3. Then there were the fixed stars; what were they?
4. Thunder and lightning were terrifying; were they not expressions of anger?
5. And the other weather phenomena - snow, rain, fog, etc. What caused them?
6. Earthquakes too occurred, for no apparent reason.
7. Life in all its myriad forms had to have been created by Someone or Something.
8. How did an acorn *know* it had to produce an oak tree and a conker a horse chestnut?
9. Then how did eggs function?

All these phenomena, and many more, would have seemed to me like miracles requiring explanation, and so the postulate of a causative God (although not necessarily a causative *and loving* God) would have seemed to me to be reasonable. But now we know how these things really happen, and no Supreme Being is directly involved (although doubtless it will be argued that He is supervising the whole business) so that although there are still a great many unexplained phenomena, I have little doubt that advancing science will rapidly dispose of them too. So those arguments, wholly compelling in 1000 AD, failed equally compellingly for me in 2000 AD.

But that is by the way. It *is* now 2000 AD. And science is advancing at an unbelievable rate and, as a professional scientist, I set myself the task of applying the famous Scientific Method to the question:

Does the God of the Christian tradition exist?

Please note that I am NOT disputing the existence of God; I am disputing whether, if there is indeed a God, He has the main properties assigned to Him by the Church. These are:

1. He is omnipotent,

2. He is omniscient,
3. He is perfect,
4. He is just and yet merciful,
5. He loves everybody who was, is, or ever will be.

(From a famous remark attributed to Jesus concerning a sparrow falling to the ground, one might also conclude that Almighty God also loves all living creatures equally, but here my theology is weak and I know not the answer.)

So I say again, I am not disputing His existence; I dispute whether, if He does exist, these properties may be correctly keep assigned to Him, and as I shall refer to them from time to time throughout this essay, I shall give them a name for the nonce: I shall call them *the five cardinal properties* of God. If there is a God, does He possess the five cardinal properties?

I am sure that some convinced believers will dispute my right even to pose the question whether their God possesses these cardinal properties; they will see it as blasphemous even to raise the question, so let me dispose of that point first. Obviously there is no blasphemy either if there is no God or if, given He does exist, He is not possessed off the cardinal properties. There remains the case (espoused of course by the Church) that He does exist and He possesses these properties. I must argue that I am not sinful in disputing this - and if I am proven wrong, maybe one day I shall find myself on the Day of Judgment defending myself in His Presence. (Even a near-atheist such as I must experience this small soupçon of fear at the thought!) *

**I'm pretty sure I should feel the need to use the second person singular and some of the now outmoded inflexions of our beautiful language as it once was - perhaps I ought to mug up on Chaucer's English just in case!*

My argument would run somewhat as follows, I think:

Lord, Thou madest me as I am. Thou gavest me that which on Earth passeth for a high intelligence and I base my case on the parable of the talents. The talent I had, the brain which Thou gavest me, I used as I presumed was Thy intent - and the majority of my life was spent as a professional problem-solver. Thy one Appearance on Earth thirty entire lifespans previously was hardly compelling evidence to convince me of what would otherwise seem to me to be a wholly untenable hypothesis. Wouldst Thou that I should not use that gift of intelligence at all times and on all occasions during my life?

and a just and loving God could hardly object. I could imagine the rejoinder:

BUT YOU GOT IT WRONG, DENNIS, DIDN'T YOU?

and that, I imagine, would be that.

From time to time, as I develop my theme, I shall come to questions whose clarification I should much appreciate, I mean from professional theologians. (I shall also list them all at the end of the text for the convenience of anyone kind enough to attempt to answer them). A pair of such questions arise now, and so here they are:

Q.1: If God exists, is it sinful to question His Existence?

Q.2: … and is it sinful to ask if He possesses all five cardinal properties?

There is yet one more point that warrants a mention. It is the assertion in some quarters (and I trust it is not the view of our Established Church) that there is merit in believing the tenets of religion blindly and without dispute (because this certainly was the view of the Church of Rome in times past). To me that is preposterous. You might just as well say that there is merit in believing that the moon is made of green cheese! The only merit in believing blindly devolves on the priesthood, not on the believers. It is the priesthood that profits from such blind belief, as they are kept in business, and if religious faith should die throughout the world, priests would be claiming unemployment benefits!

I'm being facetious; I certainly do not think for a moment that the majority of today's Anglican clergy propagate the Gospel for any reason other than a deep-seated belief in its truth; our local clergy are delightful and honorable men. But as a schoolboy in the 1930's, I used to read the writings of Dean Inge (the *Gloomy Dean*, he was called - he was Dean of St. Paul's) in the *Evening Standard* (the joke at the time being that Dean Inge was both the Pillar of the Church and Two Columns of the *Evening Standard*) and I vividly remember one article he wrote in which even he expressed his doubts about the truth of the religion he practiced: and you know that witch doctors, voodoo man and like do maintain their hold over their believers by trickery, so it seems to me that faith *unsupported by at least some* evidence is wholly unacceptable. There is no merit whatever in blind faith. *

**It worries me very much to see missionaries from our technologically advanced society proselytizing in the uneducated Third and Fourth Worlds. The missionary appears so much more capable then the* *ignorant inhabitants and so his words are accepted unquestioningly. His God is seen to be so much superior to the local gods. As a near-atheist I object fundamentally to this. Indeed, this also explains why today the religious are found largely among the less educated members of our Western Society, among the poor and the elderly and the deprived; I think it is just because the priest is seen to be so much better educated than they are.*

So here I go with the Scientific Method, and in this case I shall have recourse to the famous trick known as *reductio ad absurdum*. This is the device of proving some proposition by assuming its contrary to be true and then showing that this leads to a contradiction, so that as the contrary is false the original composition must be true. This message is for ever turning up in mathematics and this in two similar forms.

In the first form, in order to prove some fact A, you assume the contrary (call it not-A) and prove that this leads to an absolute contradiction; it follows that you know not-A to be false and so you know the original hypothesis (A) to be true. This is the route of the pure mathematician and it is always 100% correct.

The second form, on which I rely to justify my near-atheist standpoint, is just slightly less compelling than the first, but its usage is more widespread; it permeates the entire science of statistics from start to finish. Here it is:

In the second form, you still start off by supposing A to be false. You then investigate the consequences of this assumption, usually by some experiment or other, or perhaps, as I shall do here, by some philosophical argument. You then end up with a result of the experiment, or a conclusion from the philosophical argument which, although not totally impossible, is nevertheless remotely unlikely. But, you argue, this is the result which follows the assumptions that A is false. Next, you ask yourself if this remotely unlikely event or conclusion would follow if A were true. If it now it appears that this event (or conclusion) would be easily believable if A were true and highly unlikely if A were false, you accept this argument to be a proof of the fact that A is true with a high degree of probability.

Not all that simple? No, perhaps it isn't, so I have included an Appendix to this paper, which gives an example of each of these two forms, explained as simply as I can. Read it or not as you like.

So for my analysis of this problem I distinguish two hypotheses, of which one MUST be true and hence one MUST be false. In accordance with normal statistical convention I shall refer to these hypotheses as H_1, corresponding to A, and H_0 which corresponds to not-A. Here they are:

H_1: Either i. God does not exist, or

ii. If He does exist, He does not possess all five cardinal properties.

(This corresponds to the case I previously called A and which, I contend, is the hypothesis which explains those things which not-A cannot explain.)

H_0: There is a God as asserted by the Christian Churches and He has the five cardinal properties. (This corresponds to the not-A case I mentioned before and, I contend, is later disproved.)

So now let me test this H_0, against evidence I have assembled. I shall conclude, many pages on, I'm afraid, that H_0 is false and so I shall conclude that H_1 is true; the Christian God, possessing the five cardinal properties, does not exist.

***

Is God Sinful?

You must remember the following fact. I am assuming that H_0 is true and that there is an omnipotent, omniscient, perfect, just (but merciful) all-loving God who rules the universe. I shall attempt to destroy that belief by the use of *reductio ad absurdum*. Until I actually go into destruct mode (as the modern jargon would have it) I assume as a fact that the universe is indeed ruled by such a God. No matter that I personally no longer believe it; this is simply part of my demonstration that the religious hypothesis is flawed.

So again I ask whether the Christian God is sinful. As I write this, I am horrified at my own temerity. How dare I even think this, let alone pose the question on paper? But look at it this way. The ancients were much bemused by the question whether their God was bound by logic. As He was omnipotent, could He not confound logic if He so willed? Could He not make two plus two equal five? I gather opinion was sharply divided on the issue. It makes me think of the puzzle which was commonly propounded in my youth (I haven't heard it recently):

What happens if an irresistible force meets an immovable object?

The answer to that riddle is obvious; both cannot exist simultaneously and so the question is a mindless one. Similarly, you cannot have immutable laws of logic and an omnipotent Being. So if the ancients might ask whether their omnipotent God was truly omnipotent or whether His omnipotence was bounded by the rules of logic, may I not ask if that same supposedly all-perfect God can be capable of sin? As you will see, I shall be forced to the conclusion that He is sinful*, at least in my eyes.

**Is it the ultimate blasphemy to say that I would not be too worried to find that there really <u>is</u> a God, but that He is imperfect? I suspect that perfection is a man-made concept that does not exist because it cannot exist. To take the facile example (which I would admit to not having properly thought through)* *how can you achieve perfection if a pedophile has to be denied the satisfaction of those powerful urges which are natural to him, even if wholly repugnant to you and me? So in our system, perfection can only be attempted by frustrating the pedophile. (Quite right, I'd chuck him into the nick and throw away the key, but that is not* <u>*perfection*</u>*, is it?)*

So what is sin? This is not an empty question. The Old Testament says *Though shalt not kill.**

**It does not say (as one might have expected) "Though shalt not kill thy neighbor" and I have often wondered if indeed the prohibition against killing was originally intended to be extended to animals also; but custom assumes it applies only to other people and the point is not worth arguing here.*

And at first sight murder is seen as the ultimate wickedness. But there is a great body of opinion (which I share) and which thinks that mercy killing of a loved one who is in great distress is not a sin - and then there are situations in which the defense of oneself or of one's loved ones regretfully necessitates killing someone else. So even killing is perhaps not always a sin.

But there are some situations in which sin appears to me to be absolute. Here are four examples:

1. The sack of Jerusalem in 1099, the responsibility of Crusaders wearing the cross of Christ emblazoned on their chests.

2. Hitler's Holocaust, in which millions of Jews (God's chosen people, according to our Bible) were exterminated en masse.

3. Rebels in Sierra Leone lopping limbs off babies.

4. Two little Belgian girls raped and then locked in a cellar and left to die with neither food nor water.

And I ask how such things may be allowed to happen by an omnipotent, omniscient, perfect, just (but merciful) all-loving God.

As a small boy - perhaps I was ten- something

like this came up for discussion at home with friends of the family and I proffered an explanation. I doubt whether at that age I could have made my point as cogently as I can now at age 79, but in effect I said this:

Perhaps God has His own frame of reference, such that from His standpoint those things we see as wicked are really not so terrible after all. For example, a little child cries heartbrokenly for a few seconds, but its parents don't bother themselves unduly, for within a minute the kiddie will be playing happily with his or her toys and all its troubles will be forgotten. Similarly, we may suppose that God knows that what we think of as suffering is really trivial and so He doesn't bother His head about it.

My argument was laughed out of court, of course, because no philosophical remark from a ten-year-old was worth consideration in 1931 (or today either, for that matter) and I doubt whether the grown-ups present even understood what I was trying to say. But the point has stayed with me all my life and for a long time I had no real answer to it. If God does not think it is a sin, why should He interfere? But now I think I have a counter argument to my childhood thoughts. It goes like this:

According to the teachings of the Church we are here on Earth at the beginning of an eternal adventure, an adventure which will continue for ever. In order to achieve everlasting bliss, we are requested to perform to a certain standard during this minuscule time period that we are here on Earth. What sets that standard? Well, it is set by all sorts of things, by Holy Writ, by the teachings of our elders (and presumed betters), by Church leaders and by our own consciences, in particular by the teaching of God Himself when he was on the Earth in human form and surely also *by that which Almighty God does and does not permit to happen on the Earth.*

And therein lies the rub. When you bring up your child you have to let him or her make his own mistakes in order that he or she will learn, and so clearly Almighty God will also let you and me make minor mistakes and commit relatively small sins in the process. But the responsible parent does not let his child play with a box of matches next to an open can of petrol, while he is filling the lawnmower! And God - a truly omnipotent and loving God - would not permit events which in my eyes constitute barbaric horrors of unimaginable savagery to occur, if He could prevent them at the figurative touch of a button, because it sends a wholly irresponsible message to you and me. It tells us that God tolerates these evils, so they are not all that bad! In a certain sense, Almighty God is conspiring to effect your damnation (and mine too) by downplaying such monstrous sins.

So here I go with my H_0 and H_1 business. If you accept H_0 you must accept that this loving God effectively connives at unspeakable wickedness - and there is an alternative. It is H_1, which is that a God with the five cardinal properties does not exist. So here is my third question:

Q.3: Why does God let some people commit unspeakable crimes?

Ah yes, and please don't let us have any of that *it's a mystery* cop out. Of course it would be a mystery if there were no alternative explanation, but there is an alternative. It is H_1: there is no such God!

There is yet a further argument which I can offer. It is not as strong perhaps as the foregoing, because there we saw what I can only describe as God's tolerance, if not active support, of monstrous human sin and in the present cases human sin is not involved. But what about natural disasters? Here are a few:

1. The epidemics of Bubonic plague in the Middle Ages.
2. The flu epidemic after the First World War that killed more people than the war did.
3. The AIDS epidemic, particularly in Africa.
4. Earthquakes around the world.
5. Famines in Africa.
6. Floods in Mozambique and Bangladesh.

All this misery could be avoided by an omnipotent and omniscient God in a second of time. If I had a cure for AIDS and I didn't disclose it, I would be a monstrous sinner, would I not? But God has a cure, hasn't He, which He hasn't disclosed? And what conclusion do you draw from that? Remember, we are supposed to be here on Earth to be prepared for an eternity of bliss if we are (relatively) sinless. What hope is there for children who are born with AIDS, or who are deprived of their parents by it? Remember the argument I advanced is a 10 -year-old which I think is perhaps more cogent here? You may perhaps argue that from God's perspective these "tragedies" are not very grave. But hold on - Life on the Earth is hardly unimportant in God's eyes is He is going to judge us for all eternity based on a few short years here, so surely there ought to be some sort of a level playing field, ought there not? And how can such underprivileged children be fairly expected to lead relatively blameless lives.

And while I am on the subject of level playing field, I was born to loving parents who helped me and encouraged me all their lives, while others have a childhood of horrors neither you nor I could imagine. Is that fair? And what about the poor unfortunates who spend all their lives in abject poverty and misery, or those who live their lives quite monstrously disabled? Because I see it as just the luck of the draw. So where does this loving God come in? Did you know that there is a philosophy which tries to "prove" reincarnation, because otherwise God is monstrously unfair? I think it's rubbish - I think it just proves that God, if He exists, is like Rhett Butler; He just doesn't give a damn.

As I said before, this argument is not as powerful as the previous one *on its own*, because there is no human intervention, so I can't counter the God-moves-in-a-mysterious-way argument directly, but you must remember that I am not advancing it on its own; I proffer it as an adjunct only to cases in which the prime sinner is clearly human. On its own the present argument may not be convincing, but I doubt if anyone would deny that it does at least *reinforce* the previous one. Moreover, there is a further point I must make. There is a whole statistical approach* to the way in which disasters occur which seems to function well enough.

**If you are interested you could investigate either extreme value statistics for less controversially perhaps, the Poisson distribution.*

The point about this is that it implies that disasters occur as a strictly random process, and the implication of that is, that the Hand of God plays no part in it. H_0 implies the interference of God; H_1 implies that a random process is at work and so does the statistical analysis. So I ask:

Q.4: Why does God let monstrous natural disasters happen?

Q.5: Why does he treat some of us so extremely unfairly?

How holy is the holy Church, anyway?

Do I remember correctly something in the marriage service about Holy Matrimony (which incidentally after 53 years of a very happy marriage I would never knock) symbolizing the mystical union between Christ and the Church? Because I for one would not wish to be associated with some of the things that Holy Church has done through the years. Why, I should like to know, should Almighty God let his Church commit some of the most abominable crimes the world has ever seen and just as surprising, why has Holy Church got so many things wrong and perhaps even more surprising, practically *never* managed to get things right? So let me examine some of these sins.

I suppose everyone starts with the Inquisition. They delivered people to the "secular power" in the full knowledge that they, fellow Christians, were to be subjected to the most terrible death man has ever devised for man. The victims were good people, so convinced of the truth as they saw it, that they were prepared to suffer that appalling torture for the sake of the God which they, their accusers and their torturers all believed in and worshipped.

Now I can accept that some of the Inquisitors, blinkered and blind though they were, at least thought they were doing God's work - but surely many others of them were pure sadists. They can't get up to that sort of thing nowadays, but a few of today's priests do go around buggering little boys, and surely it is just a matter of scale. So I wonder how a loving God, who could've fixed things in the twinkling of an eye, did nothing about it, because even if the God-moves-in-a-mysterious-way argument of my childhood holds water regarding the victims, the *ex post facto* result of the Inquisition has been to bring God's Holy Church into grave disrepute and - stretching the point a bit further - to endanger the immortal soul of people such as I who say that a loving God therefore doesn't exist!

Then there was the dreadful sack of Jerusalem that I have already mentioned; I raise it again because it was the direct response of medieval society to the appeal of Urban II for the liberation of the Holy Land from the infidel*; and let us not forget the claim of the Church of Rome for Papal Infallibility when the Pope is speaking *ex cathedra.*

**"Infidels" who believe in the same God and venerate Jesus Christ as a prophet second only to Muhammad!*

Now God is supposedly omniscient. He knew what the consequences of Urban's call to arms would be *and yet He let things go unchecked.* Again God's supposedly Holy Church is brought into disrepute by the action of its priests (or rather, this time by its chief priest).

Then we had the Borgias, the Great Schism and I know not what else. If you add all that up, you see that the indictment present in Martin Luther's 95 articles pales, as they say, into insignificance beside the other monstrous evils. And God did nothing to protect His Church from all these accusations.

So you must know what question I am going to

put now:

Q.6: Why did God permit His Church to commit horrendous crimes?

The Doctrine of Free Will

(After I had sent this paper to several people and as a consequence of a discussion which I then had with my lovely granddaughter - who is at present reading History at Oxford - I have inserted this as an additional paragraph.)

My wife and I spent Christmas with our daughter and her family and I mentioned that I had written this paper. My aforementioned granddaughter then said that what I had said would be countered by quoting the Doctrine of Free Will: this struck me as a comment of such significance that I ought to include a rebuttal in this document. It seems to me that this is the point at which I should do that, so here it goes:

Let me start with Occam's Razor, or rather the

philosophical argument which normally is so known (for this is not exactly what William of Occam, a 14th century Franciscan friar, actually propounded*).

**What he actually wrote was Pluritas non est ponenda sine necessitate.*

I find on the Web a couple of excellent practical definitions of what is today known as Occam's Razor which fit my bill. Here they are:

1. If you have two theories which both explain the observed facts, then you should choose the simpler until more evidence comes along.
2. The explanation requiring the fewest assumptions is most likely to be correct.

So my point here is that there is no need to introduce a further doctrine, that of Free Will, to explain things. You see, you postulate a God and then you have to postulate the Doctrine of Free Will to explain His behavior - but without the God-postulate you don't need the Free-Will postulate either. So Occam's Razor would automatically reject the God-postulate here. (It is interesting to note that William of Occam got himself into hot water with Rome in those days just because of his philosophy!)

I doubt whether the foregoing argument will impress too many people, so now let me examine the doctrine of Free Will. God lets us do what we like, does He? Now I have no objection what ever to a system in which I may do what ever I wish, and consequently may go to the Devil in my own way (if the Devil exists of course) *provided I do not harm anybody else.* Indeed, I find it highly repugnant that arrogant politicians pass laws which forbid me to take drugs if I want to take them* provided I harm nobody but myself.

**I don't want to take drugs, but that is not the point; the point is that no man has the right to stop me from taking drugs if I want to.*

I also find it repugnant that I am compelled by law to

wear seatbelts when driving my car, although indeed I have been wearing a seatbelt since they were first introduced - but that is my choice. No man has, I think, the right to force me to wear a seatbelt.

The situation is wholly different, however, if my actions harm or even tend to harm, others. All societies, at all times, have had laws which tried to prevent one person from harming another and which impose sanctions to enforce this. But not God? He could easily give everyone Free Will, provided no one else was injured. Does the Church really assert that their God cannot devise a system whereby Free Will is trammeled somewhat where necessary? Can't He do everything? Isn't He omnipotent?

What do you know of Aleister Crowley (1875 - 1947)? He was known as the "Wickedest man in the World." He was a homosexual pedophile of almost unspeakable depravity; for example, he had young virgins raped in front of his altar - and he is famous for his dictum

Do what thou wilt be the whole of the law.

Now I cannot see that this means anything other then that the Doctrine of Free Will applies to all men, so I come to my next question:

Q.7: Does the Doctrine of Free Will not therefore equate God to Aleister Crowley?

The illogical absurdity of this sticks in my gullet. Am I to believe in an omnipotent, omniscient all-loving God who permits unspeakable horrors to occur, because He cannot device a scheme for preventing them?

Inner Conviction

I shall be told that none of my arguments hold water, because my opponent has some form of *inner conviction* that he is right and so that I am wrong. No argument will shake this belief, he will say, because he *knows* he is right. Logic is irrelevant in the face of this "knowledge."

I hold beliefs too. I have faith in my wife, my children, the things they do and say. If you ask me I say I *know* these things. But if I were to be given sufficient evidence to the contrary - say if my wife attacked me with a carving knife one day - I should have to reconsider my position. There are very few things one can really know in the literal sense, are there not? So let us examine this business of "inner conviction."

I spoke to a charming lady the other day who was collecting for the Salvation Army. I gave her small sum and we chatted. I said I denied her Faith and she said that she didn't want believe that there is nothing more to this life then a short stay on the Earth and then oblivion. I agreed, but I said that although I don't like the idea, that doesn't mean it's wrong.

When I was a young man we had a war. I didn't like that either but there was a war nevertheless. My sister-in-law once told me I should believe in God because "it is so comforting!" I was speechless to think that any rational human being could so put the cart before the horse. But she did.

That some people have such convictions is not up for dispute; the lady collecting for the Salvation Army was an example, but the question remains whether these convictions have any real significance if there is little or no other supporting evidence. Those members of the Roman clergy who *genuinely* held that Galileo was wrong would no doubt have sought to justify their stand against him by appealing to their inner conviction that the Earth was the center of the universe, look how horribly wrong they would have been! The human brain can play awful tricks and one such trick is a form of self-hypnosis, so that people can convince themselves that they believe what they want to believe. I therefore think it is unreasonable to quote one's inner conviction in the face of a mass of contradictory evidence; but that is just what I happen to think and the innerly convinced may not accept that argument. After all, the Jesuits always claimed that if they had the control of a child until it was five or six years old, they had control of it for life!

I must therefore look elsewhere if I am to raise doubts in their mind. Where should I look? I think I can present a well reasoned case:

Human beings have a huge range of different reactions to religious ideas. Some accept them readily - indeed, so readily that even the most convinced evangelist must worry at the unquestioning way some respond to newly proposed dogma. Others are such as I. I would take a whole wealth of convincing that any religious concept was worth a moment's consideration, let alone adoption. To choose a pair or fairly aggressive adjectives, men and women range from the foolishly gullible at one end of the spectrum to the horribly cynical at the other.

Now I think you would agree that people who based their argument on "inner conviction" can hardly be drawn from my "horribly cynical" end of the spectrum, and although I do not actually label this as "gullible" I think it is fair to argue that they come in the main from what I shall call the *gullible half* of the spectrum. Let me assume at this point, gentle reader, that our common friend George is to be found in this rather pejoratively labeled *gullible half.* George is a fine chap, but perhaps he is easily persuaded. He certainly claims to know that the Christian religion is correct and cites "inner conviction" as his justification for this review.

Now suppose George had been born of Jewish parents and had been named Isaac. Would Isaac not usually be of the Jewish Faith? Please note, I am not knocking the Jewish faith and more than any other. I am just pointing out that Isaac would no doubt adhere as strongly to the Jewish Faith as George does to the Christian. Isaac too would claim an inner conviction - but would reject Jesus Christ as the Messiah, would he not? And if George's parents had been Muslims and had named him Ali? Ali would have believed that every single word of the Koran is literally the word of Allah; indeed, his main education would have been to learn the Koran word for word by heart because of that fact, and *he would have believed that also from inner conviction.* And if he'd been born of Indian parents? Or Japanese? Or into one of the other countless religions to which people subscribe worldwide? Would he not nevertheless have claimed "inner conviction" for his religious faith?

It follows then, does it is not, that the particular "inner conviction" of which you are seized is an accident of birth? So are eternal verities based *only* on "inner conviction?"

So now as I phrase my questions concerning this part of my thesis, I take it as a given fact (as an axiom of my philosophy, if you will) that nearly every religious person subscribes to their religion into which he or she was born*.

**The non-religious such as I, naturally reject all religion. But I must confess that the effect of my birth and infant nurture is that I only seek to deny Christianity, the Faith, into which I happen to have been born. Judaism, Islam, Buddhism I respect as I do other faiths, but also automatically I see them as in error, and so I see no need to deny them. For me they are wrong ab initio; and is that not evidence of the effects of my origins?*

There are a few people who change Faith, of course. You get mixed marriages in which one party changes her or his *avowed* Faith in order to marry a partner of the other. I wonder cynically how deep such a change of Faith actually runs, so I don't think that really counts. Then there are people who move between one branch of a Faith and another - Anglicans who embrace the Church of Rome for example - but these are relatively minor changes in belief. No, I am concerned with people who deliberately and after mature consideration abandon the Faith of their childhood and genuinely and enthusiastically embrace another Faith because of a true conviction. I suggest that they are very few in number.

If you disagree with this thesis, the following questions are NOT for you and I suggest (if you have patience to continue reading all this) you move onto the next section. Furthermore, if you do not claim *inner conviction* as is the reason for your religion, fast forward again.

But if you agree that genuine change of childhood Faith is rare, and also you claim to be moved by inner conviction, I have two questions for you to answer:

Q.8: If a deeply religious Christian had been born a Muslim, would he not as a consequence have adhered just as strongly to Islam as he now does to Christianity?

Q.9: If that is so, does this not make inner conviction highly fallible?

I am not knocking Islam, of course. Its ethical teaching, perhaps more appropriate to life in Arabia then to life in the UK, is fine. But my point is that Islam differs from Christianity and so both cannot be correct - and yet Ali subscribes to his Faith just as zealously as George does to his and both justify their belief by appeal to *inner conviction.*

Holy Writ

Critics of religion commonly point out that the Old Testament contains logical errors. I am sure there are many such errors, but as I am not a theological scholar I am ignorant of them. So here are just a couple; I ask how and why, under their religious hypothesis, such errors can exist:

1. It is notorious that Cain, having slain his brother Abel, goes off to another country and takes one of the women living there as a wife. And yet Cain was the son of the first man, Adam.

2. Noah takes a pair of each animal or bird on board his ark (of which incidentally we have the dimensions, and it wasn't all that big). Does this include Jaguars and sloths from South America, kangaroos and echidnas from Australia aardvarks from South Africa, polar bears from the Arctic and Emperor Penguins from the Antarctic, not to mention more than 1 million different classified species of the coleoptera alone?

Okay, so the standard answer is that Genesis is

really just a story, made up (by God?) to satisfy the minds of Early Man. But hold on. This is the word of God, isn't it? And if it is the Word of God, surely it is not addressed just to people living several thousand years ago, is it? Well, at the very least it can only have been addressed to such people as there are, until God gets round to publishing the second edition. So isn't it high time he came out with one? Why, I must ask, do I have to be satisfied with the Revelation of an Almighty God that is so out of date as to be almost meaningless to me and which contains glaring errors? It doesn't even have an erratum! Surely this all-powerful, all-knowing, all loving God could do a bit better than one edition every five or six millennia, couldn't He?

Then if there is a God who loves us, why doesn't he appear now and then on Earth? One appearance two thousand years ago in an obscure village for a world with an ever growing population of six billion is hardly frequent. And think of all those souls who never had a chance of being saved, because they have never been exposed to the presumed One True Faith (i.e. H_0). Is a loving God also a monstrously unfair

God?

Then there is something else. Why is there nothing in Holy Writ concerning things of which its actual Earthly authors knew nothing? I don't mean complicated modern discoveries like Heisenberg's Uncertainty Principle, or the General Theory of Relativity which nobody could then have understood. I mean things that even Early Men could have easily grasped four thousand years ago. Why for example does it not tell us that the Earth rotates on its axis every 24 hours, or that it goes round the sun? Could we not have been given just a little medical advice, say concerning the role of flees on black rats spreading that bubonic plague? And what about a little mathematics - it would have been most helpful. *As a loving parent I gave such information to my children at an early age.* Why didn't God do that for His children?

Perhaps worse than that, there is some scientific rubbish in the Bible, isn't there, and I exclude miracles here? For example, people are "possessed of devils," are they not? Why not explain a little bit about psychiatry? Why perpetrate meaningless myths when the truth is easily explained? You see that all these problems disappear once we admit this so-called Holy Writ was indeed Human Writ.

Now look at the scientific advances since Christ was on Earth. Mathematics, astronomy, physics, the biological sciences, all of these things have advanced without appeal to religion, without any Divine Intervention. The world is changed unrecognizably, but Holy Church still relies on the Holy Writ. What has Holy Writ contributed to the advancement of knowledge? It seems to me that all Holy Church and Holy Writ have done has been to deny progress and retreat as science has produced more and more evidence that the Church was postulating a load of rubbish.*

**Napoleon teased Pierre Laplace that there was no mention of the Creator in his treatise Méchanique Céleste. Laplace's reply is famous and fully authenticated. He said: Sire, je n'avais pas besoin de cet hypothèse-la.*

A load of rubbish? Well, the Earth was supposed to be the center of the universe, and when Galileo ventured to suggest the opposite they clobbered him. Then Darwin came up with his theory on *The Origin of Species* and the Church attacked him - and lost. (And the Bible Belt in the States still goes on insisting Darwin was wrong and schools must teach "Creationism" or what ever that rubbish is called.)

So why doesn't God disabuse them? Of all the advancements science has made in the last two thousand years, I challenge anyone to produce a single example attributable to religious thought. So my next question is:

Q. 10: What evidence is there that God had any hand in the production of Holy Writ?

You know what I believe, of course. Holy Writ is as it is (and here I include the Holy Writ of *all* religions) because none of it was written by God, nor was it written under the supervision of God or even with the tacit consent of God because, as I said before, if He exists He couldn't care less. Holy Writ was written solely by Early Man; for proof of this, please see the Holy Writ that early Man wrote.

The Human Soul

As far as I understand it, I am conceived of as possessing an essential non-corporeal element which is the real I and which continues to exist after my bodily death. This is called the soul and its future course (after my death) maybe be benign, in which case it goes to Heaven, or malign, in which case it suffers punishment of some form or another, presumably proportionate to the sins I committed on Earth. Assuming I have been relatively sinful, it is unclear to me whether I, in the form of my supposedly immortal soul, can ever end up in Heaven. But that is by the way; I want to examine this soul-concept as I understand it.

For me the first stumbling block is this nebulous *soul* itself. I imagine it is supposed to be an appendage to my body which is supposed to represent the inner *me,* the thinking, feeling, emoting self-aware being that I believe I am. Remember Descartes? *Je pense, donc je suis.* *

**If you think he said cogito ergo sum you would not be actually wrong, but he thought it in French first - and published it before he rewrote his philosophy in Latin some years later. So there!*

So the Church postulates the existence of an otherwise undetectable body, or substance, which has no apparent physical properties except those it is convenient to assigned to it. Now where have I met this concept before? I can think of three examples of old-fashioned physics:

There was phlogiston, which was released when things burned; it was weightless, as I remember and otherwise undetectable. Then we had caloric. That was another indivisible, undetectable principal which was involved in the transfer of heat. Quite recently we have had the luminiferous ether; that was supposed to be the medium that transmitted electro magnetic waves, and again it was otherwise undetectable. They were all hypotheses and they have all been discarded. So why should we believe in the soul, for whose existence there is no external evidence?

What are we trying to explain by means of this postulate? Fifty years ago I too accepted that I had a soul; I identified it in my own mind with my individuality, my personality, my own uniqueness. Well, I still accept that I am a unique individual with my own personality, but advances in computing force me to accept that I am nothing other than a rather elaborate (and not terribly accurate) computer. My erstwhile soul is now for me nothing but a representation of the (highly individualistic) wiring diagram which represents my brain. Now that I know how computers work, I know (in principle at any rate) how my own brain functions, and I realize that I am nothing more than a walking, talking computer, an electro-chemical machine. For me the only mystery left, apart from a load of problems of detail, is how it comes about that I am a self-monitoring computer. I am self-aware, and what that means I don't really understand. But there isn't much left of the concept of furnishing me with a soul; I think my self-awareness is just a consequence of the way my brain happens to be wired up! This isn't a proof in the sense of the H_0 vs H_1 argument, of course. All I do is to proffer an alternative to the hypothecated soul. So my next question is:

Q.11: Why postulate the soul when computer-theory, only a few years old, now explains nearly all there remains to be understood?

It seems to me that the existence of a soul has to be postulated in order to have a basis for another hypothesis, which is that I shall survive my own death. Obviously my body doesn't survive my death, so if you want to hypothesize that death is not final, you have to invent something which does not die. This, I presume is the reason for asserting the existence of the human soul.

So there you are, believe in the existence of your immortal soul and death suddenly stops being the end! I find it a most attractive theory, but unfortunately the fact that a theory is attractive doesn't make it correct. When my lovely sister-in-law told me "believing in life after death is comforting," the devastating illogic of that remark has haunted me ever since - but of course many people will believe in life after death, if you tell them it is true, the idea is so consoling. So let me examine the evidence:

When the theory of the immortal soul was first mooted, it had not been recognized that *Homo Sapiens* is just an animal, as other animals are. *

**How it can have escaped the notice of the Ancients that we are also animals, amazes me. For example, all of mammalian reproduction is pretty well identical between man and his domestic animals - and they were known to have hearts and lungs and kidneys and so on, just as we have. The Ancients must have been blinded by the superior position Man had given himself. They call it hubris. To me the fact that I am an animal is blindingly obvious.*

It was then tenable, I suppose, to postulate that Man differed from all other living beings in possessing a soul. After all, Man possesses a rational, thinking brain and it was thought that animals do not: but if we examine those ideas a little more closely, all sorts of additional problems arise in the light of modern scientific discoveries, and these you must explain if the theory of the soul is to have any possible credence. Here are some of them:

First of all, it is now abundantly clear that many animal species do possess rational, thinking brains. Chimpanzees do; they can be taught sign language and once taught, they pass the language on to other chimpanzees. They are capable of solving simple problems. *

Here is an example. You hang a bunch of bananas from the roof of the cage and give the chimpanzee a stick that is too short to reach the bananas. The cage also contains a fairly heavy crate in one corner. The chimpanzee drags the heavy crate across the floor of the cage and positions it under the bananas; then he gets on the crate and gets at the fruit using this stick. That is pure reasoning, pure (albeit simple) problem-solving. I wonder if the average three-year-old child would think of doing it.

Again, different groups of chimpanzees have different skills - for example, only some groups know how to crack heavy nuts open by using two stones as hammer and anvil, while if one member of the group can do something, so can the others, so you see this is an elementary form of differing chimpanzee cultures. Clearly chimpanzees think, so Man is not unique in thinking. And although Moses didn't know it, *we* now know that Man and chimpanzee have 97% of their DNA in common; so how far do we really differ from the chimp?

Then there was the famous case a few years ago when and little boy fell into the gorillas' enclosure in a zoo. I think Whipsnade and a large male gorilla mounted guard over it.

Indeed, I would argue that dogs also think in a simple way. Our family had an exceptionally intelligent dog who could understand a new command uttered just once and then obey it thereafter. Of course this is only subjective evidence, but it is as good as much of the evidence offered by the religious.

And now Charles Darwin intervenes, and again we get an argument that has only emerged long after religious dogma was promulgated. If we trace the descent of Man backwards in time, we go from *homo sapiens* roughly back to *homo erectus*, then to *homo habilis,* on to *australopithecus* and so on, back and back until we finally end up with some primeval bacterium at the dawn of time. Now I don't suppose anyone seriously suggests that any of the primeval bacteria which were my ancestors,* possessed immortal souls, and yet unknown though it of course is, there must be a direct line of descent (or rather, a multiplicity of direct lines of descent) leading from bacterial life in the Pre-Cambrian to me in the year 2000 A.D.

**It's difficult even to say who were the parents of*

whom, when talking of bacteria, for they multiply by simple cell division; at that stage in the evolution of life there were neither parents nor children.

To fix our thoughts a little more, suppose we go back no further in time then to the commencement of sexual reproduction, starting presumably from some marine invertebrate, perhaps in the Cambrian some five or 600 million years ago. Mind-boggling though you may find the thought, there has been one direct descent from one Cambrian creature, through untold millions of generations and always in the male line, to me. That was one unique sequence of male animals; and if this sequence starts off with animals without a soul and if I have a soul, there must have been a point in that sequence at which soulless animals fathered one which had a soul. (I cannot see how God's purpose can be obtained by some sort of gradual development of the soul. It must be a yes or no situation; either you have or you have not.)

So if we combine Darwin with the teachings of Holy Church we are forced to the conclusion that

God intervened at some point in the development of life and picked out of those who were to begin the process of having souls. This is a direct consequence of the Darwinian theory; no wonder the backward groups in the USA wish to retain Creationism; Darwinism and the concept of a soul are simply incompatible.

There is another very real problem in all of this. Let us go back to the first in my long line of male ancestors who had a soul, and imagine his arriving at the Pearly Gates. They let him in and he starts wandering around Paradise, looking for his father (of whom he was particularly fond). No Daddy! So he goes up to one of the angels and registers a protest. Why has my father been discriminated against? Or rather, why have I been discriminated against by being almost the only chap here who hasn't got his dad with him to go down to the local and have a couple of jars of nectar on Saturday evening? So perhaps I put it facetiously, but this wouldn't be funny; it would be downright wicked.

So here are my next questions:

Q.12: What is the evidence for the existence of the soul?

Q.13: How can there be a being with a soul whose father had no soul?

Being 'Saved'

A great deal is made of being "saved," salvation being the apparent consequence of being a Christian. I don't understand what this means, because evidently the most noble and generous people who are not Christians cannot be "saved." I think this assertion that salvation is only available to believers is so mind-bogglingly arrogant, that if being "saved" means I have to spend eternity in the company of such bigots that promulgate that idea, I could do better to spend eternity (if it is available) in the company of almost any other group of people.

And what about other people who have never been exposed to Christianity, let alone Jews and

Muslims? For example, could even the noblest of the Aztecs, the Incas, the Mayas have been "saved?" And if they have lived good, honorable lives, what happens to them when they die? And what about Pizarro, murdering all before him in the name of the Church - I suppose he was "saved," was he, because he believed in Jesus Christ? And what about babies who die in the first few days of life?*

**Naturally after they have been baptized. Clearly they could have no possible chance if a few drops of Holy Water hadn't been dabbed on them before they snuffed it!*

Because it's a bit hard on me, who has been sinning away now for nearly 80 years and (according to the Church) am in grave mortal danger, while a baby of three weeks is home and dry, provided he pops his clogs early enough.

Q.14: Is God really so uncharitable as to deny salvation to nearly everyone?

Here I must interpolate a short comment on the concept of Original Sin, again as I understand it. My lovely wife and I have two children; they were conceived and brought up in love. They are now grown up and have children of their own; they too are fine, upstanding members of society. To suggest to me that those innocent babies were somehow born in Sin and only escaped damnation (or whatever) by being baptized when they were a few weeks old, is such a disgusting travesty of the thinking of any decent, honorable person that for me, it condemns the Church out of hand. Indeed, the Roman Catholic Church was forced into the special dogma of the Immaculate Conception of Mary, the mother of Christ, because otherwise Mary would have been born in sin and she had to be perfect! To such absurdities is the Church reduced in order to maintain what I see as an untenable position.

Going back to the business of being "saved," and if I understand matters properly, I am to be "saved" if I both believe the Christian teaching and repent my sins, because Christ himself (He was God, remember) took my sins on His shoulders and "died, so that I might live." God himself was the Scapegoat for the entire world, was He?

There is a magnificent and fascinating 12 volume work written between 1890 and 1915 by Sir James G Fraser, one of the founders of modern anthropology, called *The Golden Bough.**

**I haven't read all 12 volumes, but I do own, and have avidly read, a one-volume condensation of the book. I thought it absolutely brilliant.*

In this he studies the religious behavior of Man and in particular the belief in a scapegoat, an idea which seems to permeate the mass of human thought. So the scapegoat concept is thousands of years old. That I accept. But this ancient, pagan concept seems to have been adopted by the Christian Church nearly to millennia ago and, once adopted, it seems not possible to discard it, and for me that is quite ludicrous. I cannot accept the concept that I can offload punishment for my sins onto anyone else and most certainly not onto my Judge. A situation in which I am charged and found guilty, whereupon punishment falls on my judge and not me, is so utterly ludicrous as to beggar belief. If I sin, I am responsible. Finis! End of argument. Remember, God is all-powerful. He is above all rules. So if He in His infinite wisdom wishes to pardon me, He may and He can. What part does the Crucifixion play in all of this? Does the Church seriously suggest that God, when a Man, had to suffer crucifixion in order to be able to pardon me?

Now if Christ were not God himself, there would be nothing to explain. He was a prophet; Pilate had him executed; God played no further part. End of discussion! But if you assert that Jesus Christ was also God, it seems to me that we have here a contradiction even for a committed Christian, let alone for this near-atheist pecking away with two fingers at his word processor. I reckon W.E. Henley got it right when he wrote:

It matters not how strait the gate,
How charged with punishment the scroll.

I am the master of my fate,

I am the captain of my soul.

Q.15: Why does the Church promulgate the pagan concept of a scapegoat?

Christ's Divinity

Christ did not claim to be divine (and neither did Muhammad) and the early Christians did not believe Christ was divine. I was astonished to discover quite recently that the acceptance by the Bishops of the theory that Christ was Himself divine did not occur until the Council of Nicaea in 325 AD, and even thereafter, it was hotly debated for a long while. That's almost 300 years after Christ's Ministry. It's rather as though somebody were today suddenly to pronounce John Wesley as divine; whoever did so would be laughed at, and yet today's Church accepts the decision taken by a load of bishops at Nicaea, at a time in which superstition ruled the world! That is the trouble with religion - it can't be wrong. Once you have postulated something, that something necessarily must stay true for all time. For example, look at the mess the Roman Church has got itself into over contraception - once committed there is no way they can reverse their views.

So the Christian Church, being a monotheistic religion, has to twist itself into knots trying to explain the business of Three in One and One in Three! To what purpose? Not many religions postulate a human God. I can think of the Caesars who proclaimed themselves to be divine and the Japanese Emperors who were apparently divine up to 1945. How we poured scorn on these two sets of quasi-divine beings when I was a kid! How ludicrous we found it that these human beings were also Gods! And yet we claim divinity for our prophet!* And that on the basis of a disputed, and apparently needless decision taken in 325 AD!**

**A Roman Catholic and a Jew are traveling in a train, discussing their sons. The Catholic says his son is about to be ordained as a priest. The Jew asks him*

But what then?

Well, he may become a bishop and live in a beautiful palace.

Yes, but what then?

He could perhaps become a cardinal, the Prince of the Church.

Yes, but what then?

I suppose it's just possible that he becomes Pope.

Yes, but what then?

What do you want, Jesus Christ?

Well, one of our boys was!

***Islam has more adherents than Christianity, but they saw no need to postulate the divinity of Muhammad.*

Q.16: So why was Christ divine?

Prayer

Another aspect of religion which seems to me to be utterly ludicrous is the idea of prayer, not so much because there is no evidence that it works, but because, even under the hypothesis that the Christian teaching is wholly correct, it is strictly illogical. There are two reasons why I say this:

The first reason that prayer is illogical is the essential sycophancy involved. When there were just a few Israelites involved wandering through the inhospitable desert with their God beaming down on them from some fluffy bit of cumulus, I can see that they thought He would like a bit of flattery from time to time. After all, Solomon, Saul, David *et al* no doubt responded to flattery and Jehovah no doubt was seen as a very human sort of figure. But today? Look at what we now know about the cosmos. There are roughly 10^{10} suns in our galaxy, the Milky Way alone. That is one followed by ten zero's, ten thousand million suns: 10,000,000,000 of them! Neither you nor I can conceive of such a number. And just about now our astronomical observations have become so precise, that we are beginning to detect planets circling some of these suns. So far only a relatively few have been found, but so far we can only detect hugely massive ones circling their suns from very close by. In fifty or so years we shall doubtless have established the existence of countless millions of planetary systems, and within those planetary systems there will surely be an enormous number capable of supporting life in one form or another, and quite possibly rather similar to ours.* The concept of a personal God sitting on a cloud way up there above our heads becomes somewhat more remote.

**A life form presumably requires an elaborate chemistry, and until recently the only contender for this was the carbon, hydrogen, oxygen system of which we are a part (although recently other systems have been discovered in remote places). So perhaps other life forms have evolved in other environments using systems wholly different from our own.*

But oh Lordy Lordy, things are far worse than that, for I have only been talking about our own galaxy, and how many galaxies are there? Well, it seems that there are something of the order of 10^{10} galaxies - that's ten thousand million of them also -

and all (as far as anyone knows) roughly like ours. That means we have to think of 10^{20}, or 100,000,000,000,000,000,000 individual suns with the goodly proportion of them having planets around them, presumably many loaded down with carbon, hydrogen and oxygen and so presumably with a great many of them having life something like ours aboard. And then there are the other possible chemistries!

Perhaps a million million million planets bearing some sort of life, and you think flattery from one of them is important to the Brain that could conceive all that?

Let me emphasize my point here. I do not believe that there is such a Brain organizing the entire universe; I believe it is all just happenstance. But the religious hypothesis holds that He does exist, and I ask whether praise, flattery, sycophancy in any of its forms can be of importance to such as He? Would you, gentle reader, pay much heed if a load of greenfly that inhabit your favorite rosebush suddenly all bowed down in homage to you? You'd be amazed that they could, of course, but would you be elated, overjoyed, delighted, proud that greenfly worshipped you? Could you imagine yourself actually *requiring* sycophantic admiration, praise, adoration and so on from your greenfly? So I say that sycophantic praise is ridiculous. I just don't see it.

Q.17: Why does God require sycophantic praise from man?

The other point is the use of prayer when asking for some sort of favor. People pray for rain, or for the life of a loved one, and armies of course pray for victory over their enemies.* God is supposed to know everything and to love us with unbounded love - so he necessarily does that which is best for us always.

**In the First World War both armies, British and German, used to line up on Sundays and pray for victory over the other lot, and their respective Churches supported them! Can you imagine anything more obscenely ludicrous?*

Indeed, prayers sometimes end with *Not our will but Thy will be done,* so what on Earth is the point of asking Him for anything? He knows what is best for us, long before we know it ourselves. It is wholly daft to try to influence Him.

Q.18: Why try to influence God when he knows better than we what our needs are?

And there is something else. The Church teaches us the value of prayer. I have seen Jews at the Wailing Wall, rocking to and fro as they communicated with their God; some of my Muslim friends traipse off to the mosque on Fridays to worship Allah the Merciful, the Compassionate; Sikhs and Hindus and Buddhists and members of multiple other religions all, in one way or another, plead for the intercession of one or other postulated deity to seek to further their own particular aims; and yet, what evidence is there of its effectiveness? Few things would be easier for the management scientist, than to set up a series of definitive experiments which would effectively answer the question whether prayer has any real value, but as far as I know it has never been done. I wonder why, and I have a nasty feeling that I know why. I suspect that either

1. Such experiments as have been made, have shown no particular advantage to be gained by prayer, or
2. No such experimental evidence for the effectiveness of prayer has ever been tried, because those who might have tried it were pretty certain it would fail anyway.

Cynical, aren't I? But then I do come from the cynical end of the gullibility spectrum! So here are my questions:

Q.19: Has there ever been a scientific investigation of the power of prayer?

and then

Q.20: If not, why not; and if so, what was the outcome?

Do you believe in miracles?

According to the religious hypothesis, miracles - loads of them - occurred while Jesus was preaching and then they continued after He ascended into Heaven, although now perpetrated by His saints. The strange thing about all this is that *the age of miracles has now passed*, to the extent that this little saying has actually become a catch-phrase. So did they ever happen?

First of all, other religions have their miracles too (which of course we Christians deny). I don't know too much about them, but certainly Muhammad descended directly to Heaven from the Dome of the Rock in Jerusalem and I gather that all hell broke loose* at some point in the Buddha's existence.

**Perhaps not a very felicitous metaphor.*

Now we Christians don't believe in the other miracles, do we? So why should we believe in our own? And what sort of signal does it send about our God if He sets down the immutable Laws of Physics and then lets His saints transgress them at will? This is not a cast-iron argument; I suppose God could break his own laws if He so willed. But would it be wise to do so? I doubt it.

Let us look at a few miracles:

1. Christ's Virgin Birth.
2. The Star leading the Three Wise Men to the stable in Bethlehem.
3. The Angels announcing the Birth to the shepherds.
4. Christ feeding the five thousand.
5. Christ turning water into wine.
6. Christ stilling the storm.
7. Christ raising Lazarus from the dead.
8. Christ walking on the water.
9. … and the mega miracle of His Resurrection.

Now if all nine of the above events were demonstrably true my case would begin to look remarkably thin. But what is the evidence?

Remember, there is no contemporaneous account of Christ's Life and Ministry; the earliest written record is based on the oral tradition and is dated thirty years or so after the Crucifixion. I do not doubt the integrity of the Gospel-writers; rather the reverse. I suspect that, because they were themselves convinced of the truth of what they were writing, they may have been tempted to embroider the truth a little in their favor.* It would be very understandable. And let it not be forgotten that they were writing in an age of enormous superstition and with almost no concept of scientific thought, so that everybody believed in devils, curses and magic in all forms.

**Rather as certain police officers have been known to depart from the strict paths of what is permitted in order to secure the conviction of a criminal they just <u>knew</u> was guilty.*

So let's have a look at the miracles I listed above. I start with Christ's Virgin Birth. Mary was a married woman, was she not? And relatively few married women are virgins are they? And why does Mary have to be a virgin anyway? St. Paul had a thing about sexual intercourse, didn't he? But where else does the absurd idea come from that there is anything sinful in reproducing the species? It seems to me that what is for me the greatest joy that the physical world has to offer, the physical interpretation of my love for my wife, can hardly be considered a sin by any right thinking person. And anyway, didn't God tell the animals in the Ark to "Go forth and multiply?"* So how were they to do that?

**Shortly after telling the animals to go forth and multiply, God came upon two snakes looking very unhappy. Upon His asking them what the trouble was, they said they couldn't multiply because they were adders. Some while later He saw them happily floating down a river on some wood and He asked them why they were now so cheerful. They explained that everything was now all right, because although they were only adders, they could now multiply as well; they had found some logs.*

So (along with a lot of churchmen) I have no

trouble in believing that the Virgin Birth is a myth, invented long after the Crucifixion by timid little men who were themselves afraid of sex.

Now I come to the Star leading Three Wise Men to the stable. This is preposterous. For a star to lead people across a desert, it couldn't have been more than a hundred meters or so high. The higher the star, the more difficult it is to say if you are underneath it or not. So is it suggested that the famous star was really less than a hundred meters up in the air? The ancients didn't know what stars were and had no real concept of what this entailed. And in any case, the story of the Three Wise Men contributes nothing to Christ's Ministry - but is a fine yarn to impress the ignorant mind. So I suggest that this is all the story is about.

This same applies to the Angels appearing to the shepherds. Why a few shepherds, for Heavens' sake? Why not do it properly and appear in the Forum at Rome? That would have made a far bigger impact. Again, I think it is just a legend invented either by the Gospel-writers themselves or by their (verbal) sources - not deliberately deceitfully invented, but with the desire to convince others of the Truth as they saw it.

Christ fed the five thousand with five loaves and two small fishes, did he? Well, there is no evidence there really were five thousand, but even if there were a lot of people listening, couldn't Christ have sent a couple of his mates down to McDonald's to get a few Big Macs on the QT? I don't know He did it, of course, but why not? Isn't that just as good an explanation of what the verbal tradition asserts?

And wasn't the story of turning water into wine probably based on the same trick, with Jesus popping off to the local wine shop and buying a few more flagons for His pals?

He stilled the storm, did he? They're out fishing and a storm comes up, whereupon His pals ask for his aid. He pops off a quick prayer and, not surprisingly, nothing happens. But it is a short, sharp storm and half an hour later it is much abated. Tell me, who gets the credit? Exactly, Jesus does - and by the time the Gospels get to be written, this little tale has become a major proof of His Status.

Now Lazarus. If he was in a dead faint, or a catatonic state, or something similar it was hardly difficult to raise him from the dead, was it, if he wasn't dead anyway? Could they distinguish true death in those days? There are loads of stories dating from countless years after Lazarus, of people waking up actually in their coffins. Such are well-attested facts. Isn't that easier to believe than the miracle as it is related in the Bible?

How did He walk on water? Very impressive if he did it, of course, but I remember seeing Peter Sellers in the movie *Being There* in which he played a simple minded television addict who rose, by a series of absurd misadventures and misunderstandings, to become President of the United States of America. At the close of the picture Peter Sellers did walk on water; there was an invisible underwater walkway which supported him. So couldn't Christ have been walking in shallows that others knew nothing about?

Finally, (and this even transcends the notion of the common-or-garden miracle) we come to the resurrection of Christ. Crucified on the Friday, he was "dead" by the Sunday. But many unfortunate wretches lived for a week or more on the cross, so was it perhaps just some sort of faint that He suffered, or was He in a catatonic state after two days, or did someone bribe the Roman guard, or did a (proto-Christian?) guard simply have pity on a tortured man? It seems to me that the entire Christian religion is based on this one event and for me there are possible reasonable alternative explanations.

Of course I do not know what really happened. But are the miracles sufficiently well-attested to be believable? Surely not. Imagine trying to prove these stories in a court of law!

And as I said before, *the age of miracles is past.* For years past the Roman Church has had to rely on medical miracles when it wants to canonize a saint, because there were no others. That's a fine cop-out; medical oddities are relatively common anyway and mis-diagnoses much more so - so as for today's miracles, again I consign them to the garbage can. Under H_0 God breaks his own laws; under H_1 someone has misunderstood or exaggerated.

Q.21: Why should I believe in miracles?

Why is Nature cruel?

We all recognize that nature is cruel, that wild animals lead highly stressed lives and many of them die horrible deaths, so I must ask why it is thus under the hypothesis of a God possessing the five cardinal properties? We all believe that cruelty to animals is a sin, so why is Almighty God able to get away with it? Why did He create carnivores? Why did he create this tse-tse fly to cause untold misery to many of the African ungulates? And harmful bacteria, and parasites that cause endless suffering to animals, and incidentally to Man also? It was all quite unnecessary; all the animal kingdom could have been made to be herbivores *if God had been organizing things in the first place.* Man has produced the RSPCA, and veterinary medicine, and generous-hearted souls who devote their lives to relieving the suffering of animals, so why is Man kinder to animals than the Christian God? Remember, we are taught that only Man has a soul; animals do not have that particular appendage; so that the Church teaches that once they have popped their clogs, they stay popped. So to what purpose were they designed if their life is to be finite and also very miserable? The Roman Church (I have been told) does not worry about cruelty to animals because they have no soul? Well, if that is the philosophy of the Christian Fellowship I want no part of it.

Now again, let's not have all that *It's a mystery* rubbish, because it's only a mystery if you believe God created the world. If you believe, as I do, that the world has developed by pure chance, you see at once how the carnivores came to develop; there was a niche in Nature which some animals could exploit, and so exploit they did. And that is why we have parasitic wasps (no less than 60.000 separate species) and magpies taking baby chicks out of their nests and cuckoos parasitizing other nests and disgusting parasites burrowing their way through human flesh, and organisms causing leprosy and syphilis and tuberculosis and so on.

It always comes down to this, doesn't it? Whatever aspect of the religious hypothesis you examine, once you probe more and more deeply into it, the more inconsistencies you come up with. There is an analogy with an interesting mathematical theorem which I remember from my student days, which says that if you accept just *one* theorem that is false, any and every other false theorem can be proved to be true. I no longer remember the proof of this but it was relatively simple. The analogy with religion is pretty obvious; get to one thing wrong, make one wrong assumption and the whole pack of cards collapses in a heap.

Q.22: Why did God design a world full of

cruelty?

What about the inaccessible world?

The point I am about to raise does not carry the conviction of the foregoing but as I have pointed out elsewhere, all my arguments are not necessarily convincing on their own. Rather it is the weight of all of them combined which has led me to my near-atheism. So I ask why God created what I call the *in accessible world*?

Fifty years ago no one knew that this inaccessible world existed. The vague hints begin to appear that life was to be found in places where Man could hardly ever be expected to be, places hopelessly inaccessible to people thousands of years ago. Remember, God has yet to publish the second edition of His Bible, so all we have to go on is what was written all that a while ago. So why did he create life forms which have absolutely no bearing on Man's existence?

I base my arguments on a brilliant book of genetics I have just finished reading called *Almost like a Whale* by Steve Jones who is Professor of Genetics at UCL (University College, London). It is subtitled *The Origin of Species Updated* and is published by Doubleday; I received it as a present from my lovely daughter in law. I couldn't put it down. Here are some quotations:

Living beings were once divided into five kingdoms of more or less equal size (animals, plans, fungi, protozoa - such as the familiar amoeba - and bacteria.) Bacteria were out on somewhat of a limb, as their genes are not contained in a cell nucleus. They seemed otherwise not much more distinct than other kinds, than plants were from animals. Now, a radical new logic has emerged. The genes show that plants and animals lie close together. Mushrooms deserve a branch of their own, closer to animals than to plants. Most of the tree belongs not to the Lords of creation, or even mushrooms, but to the bacteria and their previously unrecognized relatives. They put human kind in its place, near bananas.

The difference between men and mice, or mice and plans is, relative to the gulf separating all three from other organisms such as bacteria, tiny. The familiar branches of existence (on one of which men, mice and plants reside) are in reality unimportant twigs... Squash a fly and thousands of microbes unknown to science will be squeezed from its gut... a handful of suburban soil contains a myriad of undescribed bacteria with genes quite distinct from the plants and animals around them.

Such minute beings are a diverse and remarkable group. Species thought to be quite similar (such as the agents of botulism and of anthrax) are, the genes show, as distinct as men and maize...

Nature's new pedigree has three great domains. One encompasses organisms whose genes are contained within a nuclear membrane... The bacteria and - quite new - the Archaea, tiny entities once seen as a mere subdivision of bacteria but in fact distinct, are each in a class of their own. Their dominion is divided not into a mere five kingdoms, but into dozens.

Not just animals, plants, mushrooms, protozoa and bacteria, but dozens of kingdoms! Then there are the bizarre creatures which live deep in the sea:

... our understanding of the deep oceans - and of the interior of the Earth itself - has undergone a sea-change. A strange and rich universe stretches far beneath our feet.

Around the ocean vents... is a new world... home to many bizarre beings. They include bright red tubeworms twelve feet long. The tentacles of these... absorb hydrogen sulfide from the hot water, and forward it to dense masses of bacteria within their own bodies...

One of the vent Achaeans lives ten ten thousand feet down... Its genes code for seventeen hundred proteins, most of which have no known matches elsewhere. It needs only inorganic compounds to stay alive, and makes methane as waste.

Other Achaeans turn up just as deep but in the icy cold, as mats on the bones of drowned whales...

And then there are other organisms which inhabit the Earth at depth - one apparently eats concrete! I ask why God created them, for I see little reason. A mystery again? Well perhaps, but again my H_1 would say that life occurs by chance whenever the chemistry permits it, and that is all there is to it.

Q.23: Why did God create life forms in immensely remote regions?

Does God make mistakes?

Here is the point which I have not yet thought through with sufficient energy to be able to quote many examples, but certainly from my point of view there is considerable evidence that God has got things wrong. *

**Please remember that this is my H_0, in which I assume the truth of the hypothesis I am attempting to disprove; so here I am assuming the existence of the Christian God only to (try to) disprove it.*

We are taught that God made Man in His own Image. This causes me to ask just what this can mean. Has God digestive organs, an appendix, heart, lungs and genital organs (not to mention a belly-button)? Okay, so I shall be told that the statement was made for consumption by Early Man and is not to be taken literally today. I was also told as a child that God could adopt human form if He so wished, but again I think this is just a cop-out and in the absence of the second edition it is further evidence that the Bible is the work of Man and not the work of God.

So regardless of whether the statement that Man is made in the image of God is literally true, look at the various cock-ups in Man's construction:

1. Why is Man physically feeble compared with other animals of the same size?
2. Why is the spine so poorly engineered?
3. Why has Man a vermiform appendix?
4. Why is the majority of Man's DNA without meaning?
5. Why does the genetic code go wrong so very often?

6. Why do 99% of all people get arthritis as their joints wear out when they age?
7. Why is Man's eyesight much inferior to the eagle's?

And I'm sure any competent physiologist could go on and on listing questions such as these almost endlessly. As far as I can see there is no answer under H_0 (the existence of the Christian God). Under H_1 however, my view that all is happenstance, there is no problem. I believe that Man evolved in the manner of Darwin's theory *with no Divine Intervention* at all - because if you claim Divine intervention you must agree that He made a pretty poor job of it!

Q.24: Why is Man so poorly engineered?

And finally… J'accuse

Here is where I become really unkind and also where, what I am going to say, becomes very controversial because it is much less clear-cut; it is just a matter of opinion. But it has to be said.

I believe that the religion (as distinct from the ethics which which it is invariably bound up) is intrinsically wicked. My belief is that priests, rabbis, imams, indeed all religious leaders are themselves sinners by the very fact that they are what they are!

There, I have said it. Now of course, I do not deny the right of anyone to believe whatever they wish. If you want to believe that the world is flat, or the moon is made of green cheese or that there are fairies at the bottom of your garden, go ahead; it's no skin off my nose. What is more, you may tell other people what you believe, because I also believe in free speech. But what you may not do is to lie about these things. To tell lies is wicked, and the more important the subject about which you are lying, the worse the sin. Just as it is obligatory for the prosecution in a criminal trial not to suppress evidence in favor of the defense, so I contend that *honest* priests should always expound the reasons some people do not accept their teachings. Having done that, priests may of course then explain why nevertheless they hold the view they do. That would be fair enough. But just to state as facts those things which are at least dubious is, I think, rank dishonesty.

I was at the funeral service of an old friend who died recently and I was shocked at some words in the burial service, as indeed I have been on several occasions in the past, by a phrase such as (and I cannot be sure that I quote it precisely):

... *the sure and certain knowledge of...*

and then it could have been *eternal life* or *resurrection* or something similar. "Just like Lady Macbeth," I said to myself, "me thinks the lady doth protest too much."

Let me restate my complaint, because perhaps it is not philosophically all that simple to follow. The Church may believe sincerely in Life after Death, and so may the priest, but you must concede that at least my arguments have a point. You may reject them, but you cannot deny that they exist and are cogent. It is therefore dishonest to use phrases such as "sure and certain knowledge" because they are simply false and tend to deceive the unsophisticated mind. The best you can honestly say is to hope for eternal life or whatever it is. The ignorant are deceived by the assertion *as a fact* of something which even eminent churchmen accept as doubtful. I assert that the use of "sure and certain knowledge" is a deceit.

Q.25: Is it not deceitful to claim *sure and certain knowledge* of Life and after Death?

The next area where I believe religion (and now I refer to all religions and not just to Christianity) is bad* of itself, is where it obtrudes into my existence to my disadvantage, where it restricts my behavior in a way which I resent.

**I had originally used the word sinful here, but on reflection that is perhaps too strong.*

This happens whenever religious politicians let themselves be swayed by their religious convictions.

For example, I believe that I may commit suicide if my life has become intolerable, say because my health has deteriorated and beyond medical help. The same applies to my wife. I also believe she should help me to die, if that is my wish and reciprocally that I should be able to help her to die peacefully, if she is beyond curing. Why not, I say? Whose life is it, anyway?

Now I know there are all sorts of non-religious arguments in this matter, and of course some wicked people would tend to abuse euthanasia if it were generously permitted; but that I am not permitted to end my own life if it has become intolerable to me because of the religious convictions of others to which I do not subscribe, and indeed which I consider quite absurd is, I contend, a monstrous abuse of power.

Q.26: What right has a religious legislator to impose his beliefs on me?

Let me quote my own daughter. She was about sixteen when she said to me one day that one of the most wicked things about religion was that people believed in it with no evidence and so they were being untrue to themselves. You can have no greater sin (she said) then being false *to yourself.* I'm not sure that is right; to deceive yourself is perhaps not as bad as deceiving someone else, but still I thought her comment an interesting one. I just record it here.

But the true wickedness of all religion (not just the Christian) is to be seen in its consequences. I don't mean tea and cakes on the vicarage lawn on a summer's afternoon; I mean Queen Mary burning protestants, and Prods murdering Micks in Ulster, and the excesses of the Taliban, and the Arab Israel wars and so on and so on. None of these people can prove they are right and yet they all *know* they are, so to prove they are right (and that God, or Allah or Jehovah is on their side) they murder the other lot. I am not a Christian, but I do try to love my fellow man. If there were no religions, there would be no religious wars. Wouldn't that be better?

So my next question follows in inevitably:

Q.27: How do you justify religious wars?

A Postscript

While I have been writing this, there has occurred an horrendous event, which I contend is wholly inexplicable under the religious hypotheses H_0. It was this.

A devoutly religious Roman Catholic couple lived on Gozo, an island part of the Maltese archipelago in the Mediterranean Sea. The lady became pregnant and in due course gave birth to a pair of Siamese twins, conjoined at the lower abdomen. According to the newspaper reports one child was effectively a valuable human being; the other was only half a person, with neither heart nor lungs and indeed with very little in the form of a brain. She was named Mary, and indeed she was little more than a huge tumor, a parasite on her sister. The parents wanted the twins left joined, believing that to be the will of God; the medical profession held that, left thus, both would surely die, but that if an operation to separate them was successful, this would kill Mary but her sister might then perhaps be able to enjoy a relatively normal life, albeit after a lot of further surgery. The case went to court, the medical view prevailed, the parents' wishes were overruled, the operation was performed, Mary of course died but as I write this, the prospects seem to be bright for the survival of the other twin. In my view (neither humble nor qualified) I think that the decision was probably the right one, and that in time the parents will probably be happy enough to have one relatively healthy daughter. But that is not the point.

According to the papers the surviving child will have to undergo multiple operations to enable her to lead a relatively normal life. Apparently she will have to have all the vital parts of the lower trunk rebuilt, more or less from scratch. Why inflict this on an innocent baby?

So I must ask why a God, possessed of the five cardinal properties, should have interfered with the normal process of procreation, so that such a dreadfully malformed birth occurred. God knew this was happening - indeed, He apparently engineered the malformation in the first place (remember, He is omniscient and omnipotent). Why? Again, I won't accept that it is a mystery. As I said before, that is a cop-out, for I say it is no mystery. It is simple, it is blindingly obvious; the Christian God just does not exist.

Q.28: Why does God permit the occurrence of monstrously deformed births?

My Questions

Let me reiterate what I have said (or at least implied) elsewhere. These are some of the problems which arise in my mind if I assume the existence of the Christian God, i.e. One Who satisfies the following conditions:

1. He is omnipotent,
2. He is omniscient,

3. He is perfect,
4. He is just and yet merciful,
5. He loves everybody who was, is, or ever will be.

The existence of such a God is the hypothesis I have called H_0. The alternative hypothesis, H_1, is that this is not so and the universe we live in, is the result of random events. (This used to be open to the objection that there has not been enough time since the world was formed to generate all the life forms there have been and now are, but this argument now fails, because recent research shows that random changes in the DNA occur very frequently and so new species can arise in relatively very short time periods.)

So here are my questions:

Q.1: If God exists, is it sinful to question His Existence?

Q.2: Is it sinful to ask if He possesses all five cardinal properties?

Q.3: Why does God let some people commit unspeakable crimes?

Q.4: Why does God let monstrous natural disasters happen?

Q.5: Why does He treat some of us so extremely unfairly?

Q.6: Why did God permit His Church to commit horrendous crimes?

Q.7: Does the Doctrine of Free Will not therefore equate God to Aleister Crowley?

Q.8: If a deeply religious Christian had been born a Muslim, would he not as a consequence have adhered just as strongly to Islam as he now does to Christianity?

Q.9: If that is so, does this not make inner conviction highly fallible?

Q.10: What evidence is there that God had any hand in the production of Holy Writ?

Q.11: Why postulate the soul when computer-theory, only a few years old, now explains nearly all there remains to be understood?

Q.12: What is the evidence for the existence of the soul?

Q.13: How can there be a being with a soul whose father had no soul?

Q.14: Is God really so uncharitable as to deny salvation to nearly everyone?

Q.15: Why does the Church promulgate the pagan concept of a scapegoat?

Q.16: So why was Christ divine?

Q.17: Why does God require sycophantic praise from men?

Q.18: Why try to influence God when he knows better than we what our needs are?

Q.19: Has there ever been a scientific investigation of the power of prayer?

Q.20: If not, why not; and if so, what was the outcome?

Q.21: Why should I believe in miracles?

Q.22: Why did God design a world full of cruelty?

Q.23: Why did God create life forms in immensely remote regions?

Q.24: Why is Man so poorly engineered?

Q.25: Is it not deceitful to claim *sure and certain knowledge* of Life and after Death?

Q.26: What right has a religious legislator to impose his beliefs on me?

Q.27: How do you justify religious wars?

Q.28: Why does God permit the occurrence of monstrously deformed births?

Two answers are wholly unacceptable to the intelligent man. They are:

1. "Inner conviction." You may have it, but I sure as hell haven't.

2. "It's a mystery." I don't accept *twenty-eight* mysteries one after the other.

***

Appendix

Reductio ad Absurdum

Example 1

As I said in the main text, there are two forms which this philosophical argument can take. I start with the mathematical form which produces a 100%-watertight proof; this example is one which was used by the ancients.

As you know, a *prime* number is a number which has no factors other than 1 and itself; the first half-a-dozen primes are 1,2,3,5,7,11. A number that is not prime, such as 10 (with factors 2 and 5) is called *composite* and what I am about to prove is that there is an infinite number of primes. Put it another way, there is no largest prime number.

To prove this, we assume the contrary - we suppose there is a largest prime and we call it *P*. Now we imagine all the prime numbers from 1 up to *P* to be multiplied together in a great long string after which we add on 1, thus:

$$1 \times 2 \times 3 \times 5 \times 7 \times 11 \times \ldots \times P + 1$$

and now we look at this (huge) number. Now, if you divide this number by any one of these primes between 1 and *P* you will get a remainder of 1, so no prime number from 1 to *P* is a factor. Regarding the number itself then, there are two possibilities. Either

1. This number is itself a prime. But it is bigger than *P* and this contradicts our hypothesis that *P* is the largest prime, or
2. It is composite - but then all its prime factors must be greater than *P* and again our hypothesis is contradicted.

It follows that our original assumption was wrong and so the contrary is true, which is that there is no largest prime number.

Perhaps I am flogging a dead horse, but for

anyone who is not very mathematical at all, here is a numerical example. Suppose you were to try to say that the largest prime number is 11. (Of course we know it isn't, but numbers get so big so quickly that I have to keep to very small numbers to illustrate my point.) Then I would say to you, what about the number

$$1 \times 2 \times 3 \times 5 \times 7 \times 11 + 1,$$

which is 2310+1, or 2311? Have a look to see what factors it's got, because if you try to divide it by any of the primes up to 11, which are of course 2,3,5,7 and 11, the remainder will always be 1. In fact, it turns out that 2311 is itself a prime number, and of course it is greater than 11. That means that 11 is NOT the largest prime. But although, in this particular case the result of our little sum, 2311, is prime, this is not always so. Here is another example.

Suppose you tried to argue that 13 is the largest prime. Then just as before I would invite you to consider the number

1 x 2 x 3 x 5 x 7 x 9 x 11 x 13 + 1,

which works out to be 30031. Now again this cannot be divided exactly by any of the primes up to and including 13, but it is not itself a prime number; it is 59 x 509 - but both these numbers are themselves primes and both are necessarily bigger than 13, so 13 is again NOT the largest prime.

And whatever number you start with, things will always work out in one of these two ways, and so always delivering a prime larger than the one you started with. There is no largest prime.

Example 2

This example comes from the world of statistics and is very slightly less convincing than the foregoing. (And I mean only very slightly less convincing.)

I take my example from the world of gambling. Suppose that, in a moment of pure madness, I have gone into a gaming den somewhere in the seamier parts of Hong Kong and I am gambling against an evil-looking croupier with him spinning a coin and me betting on its falling either heads or tails. I wager five dollars a throw and after twenty spins I have lost twenty times. That's a hundred bucks I am down, and I wonder if I'm being cheated. The spins look to be fair enough, but…

Is it possible to get a heads-or-tails guess wrong twenty times in a row if the spin is fair? Obviously it is possible. But is it likely? Equally obviously it is not. So how does the statistician go about this?

Well, he says that there are two possibilities; either the coin is being spun fairly or it is not. So as he suspects that it is not being fairly spun, he starts off by assuming that the spinning is indeed fair. He expects to disprove and, by common consent, this is invariably called H_0. Then there is the contrary hypothesis, the one he expects to justify, which is that there is some hanky-panky going on; by common consent this is always called H_1. Obviously one or other of these hypotheses, H_0 or H_1, must be true and one must be false.

Now, our statistician friend starts off by assuming that H_0 is true, which is that the coin-spinning has been fair. What then, he asks himself, is the probability of losing twenty spins in a row? Now this is easy to work out. The chance of losing the first spin is obviously ½. So is the chance of losing the second spin, so that the chances of losing two in a row are ½ x ½ or ¼. To get the chance of losing three in a row you have to multiply by another ½, so that the chance of losing three in a row is ½, while the chance of losing no less than twenty in a row can be calculated as

½ x ½ x ½ x ½ x ½ x ½ x ½ x ½ x ½ x ½ x ½ x ½ x ½ x ½ x ½ x ½ x ½ x ½ x ½ x ½

and if you care to work this out, you will find it comes to a probability of 1 in 1,048,576, or just under one chance in a million.

Now here is the logic. Assume H_0 (i.e. that the

coin-spinning has been fair) and you have to accept that a one-in-a-million chance has come up. Not impossible, of course, but highly unlikely. Assume H_1 (that the ugly customer spinning the coin is pulling a fast one) and there are no problems. So you assume you're being cheated. Theoretically that is not absolutely certain, but who quibbles about that?

Summarizing this, you assume what you think you may be able to disprove. Arguing from that standpoint you come up with a result which is highly unlikely, so you say that you prefer to disbelieve the original hypothesis.

That is the form of *reductio ad absurdum* on which I rely in this little monograph.

***

Made in United States
Orlando, FL
05 November 2026

90230232R00069